2005 | Intermediate I

[BLANK PAGE]

Intermediate 1

Physics

2005 Exam

2006 Exam

2007 Exam

2008 Exam

© Scottish Qualifications Authority

First exam published in 2005.
Published by Leckie & Leckie Ltd, 3rd Floor, 4 Queen Street, Edinburgh EH2 1JE
tel: 0131 220 6831 fax: 0131 225 9987 enquiries@leckieandleckie.co.uk www.leckieandleckie.co.uk

ISBN 978-1-84372-653-1

A CIP Catalogue record for this book is available from the British Library.

Leckie & Leckie is a division of Huveaux plc.

Leckie & Leckie is grateful to the copyright holders, as credited at the back of the book, for permission to use their material.
Every effort has been made to trace the copyright holders and to obtain their permission for the use of copyright material.
Leckie & Leckie will gladly receive information enabling them to rectify any error or omission in subsequent editions.

FOR OFFICIAL USE

Total Marks

X069/101

NATIONAL
QUALIFICATIONS
2005

TUESDAY, 24 MAY
1.00 PM – 2.30 PM

PHYSICS
INTERMEDIATE 1

Fill in these boxes and read what is printed below.

Full name of centre

Town

Forename(s)

Surname

Date of birth
Day Month Year

Scottish candidate number

Number of seat

1 All questions should be answered.

2 The questions may be answered in any order but all answers must be written clearly and legibly in this book.

3 For each of questions 1–5 there is only **one** correct answer. Write down, in the space provided, the letter corresponding to the answer you think is correct.

4 For questions 6–18 write your answer where indicated by the question or in the space provided at the end of the answer book.

5 If you change your mind about your answer you may score it out and rewrite it in the space provided at the end of the answer book.

6 Before leaving the examination room you must give this book to the invigilator. If you do not, you may lose all the marks for this paper.

SCOTTISH
QUALIFICATIONS
AUTHORITY

©

[BLANK PAGE]

Page two

Marks

1. Which row gives the correct colours of the insulation on the three wires in a mains flex?

	blue	*green/yellow*	*brown*
A	live	neutral	earth
B	earth	live	neutral
C	neutral	live	earth
D	earth	neutral	live
E	neutral	earth	live

Answer ☐   1

2. Which row shows the correct symbols for both components?

	variable resistor	*fuse*
A		
B		
C		
D		
E		

Answer ☐   1

3. Which list contains input devices only?

   A   LDR, switch, buzzer
   B   LED, switch, thermistor
   C   LDR, thermistor, microphone
   D   LED, loudspeaker, lamp
   E   Motor, switch, thermistor

Answer ☐   1

[Turn over

Marks

4. An AND gate has two inputs X and Y and one output Z.

Which row gives possible logic levels for this gate?

	Input X	*Input Y*	*Output Z*
A	0	0	1
B	1	0	0
C	1	1	0
D	1	0	1
E	0	1	1

Answer ☐ 1

5. The resistance of an LDR increases as

A   the temperature increases
B   the light level increases
C   the temperature decreases
D   the light level decreases
E   the sound level increases.

Answer ☐ 1

DO NOT WRITE IN THIS MARGIN

Marks

6. The diagram shows a satellite orbiting the Earth. The Earth turns once in 24 hours. The satellite stays above the same point on the Earth's surface.

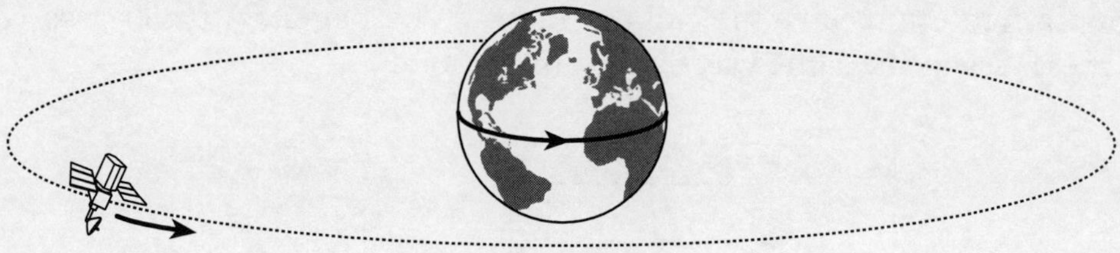

(*a*) What name is given to this type of satellite? **1**

(*b*) What is the time taken by this type of satellite to complete one orbit around the Earth? **1**

(*c*) What type of aerial is needed to receive a strong television signal from this satellite? **1**

[Turn over

Marks

7. A village fire-fighting crew is formed by local volunteers. In an emergency they have to get to the fire station quickly.

    (a) Some firefighters receive the call out message by pager. A pager receives text messages in the same way as a mobile phone.

    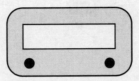

    (i) What type of signal carries the message to the pager?

    1

    (ii) What is the speed of this signal?

    1

    (b) Some firefighters receive the call out message at home by telephones connected by metal wires.

    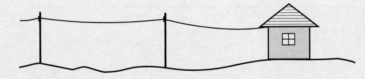

    (i) What type of signal travels in the wires?

    1

    (ii) What is the speed of this signal?

    1

DO NOT WRITE IN THIS MARGIN

Marks

7. (continued)

(*c*) One firefighter is working in the open air at a distance of 830 metres from the fire station. He hears the sound from a siren which is on the roof of the fire station. The sound takes a time of 2·5 seconds to travel from the fire station to the firefighter.

Calculate the speed of the sound from the siren. **2**

(*d*) The fire station receives the emergency signal along an optical fibre link.

(i) What type of signal travels in the optical fibre link? **1**

(ii) The diagram shows part of the optical fibre. State what happens to the signal at point X on the diagram.

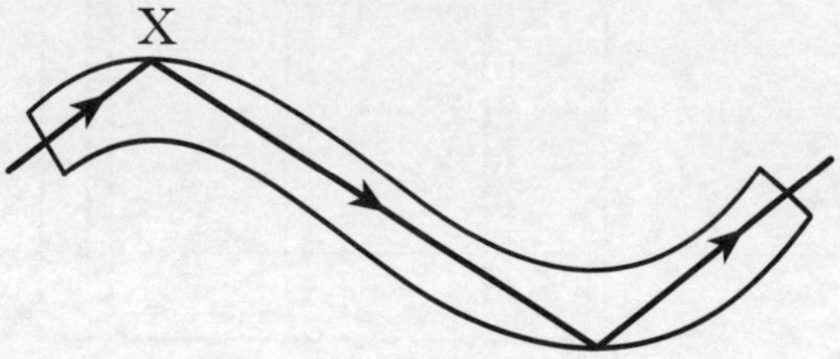

1

(iii) What is the speed of the signal in the optical fibre? **1**

Marks

8. (*a*) A computer program can produce any colour by mixing different levels of red, green and blue light on the screen. Three sliders change the brightness of each of these colours on a scale of 0 to 10.

The settings shown here produce a bright orange colour.

computer screen

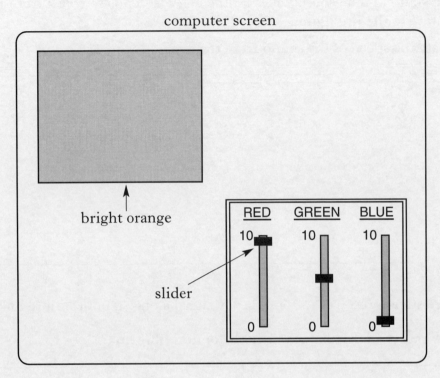

On each of the diagrams below, mark the three slider positions to produce the following.

(i) Bright White

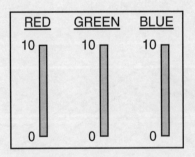

1

(ii) Bright Yellow

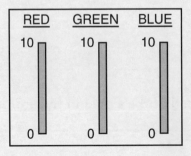

1

Marks

8. (a) (continued)

   (iii) Black

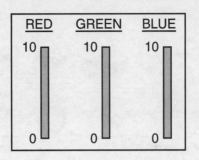

1

   (b) A computer monitor contains a picture tube which is similar to one in a television set. The computer sends a signal to the monitor.

   (i) The diagram below shows the monitor system. Part X is not named.

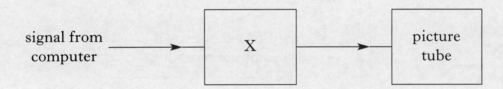

   Which of the following is part X? **Circle your answer**.

   aerial   amplifier   tuner   loudspeaker   1

   (ii) What is the purpose of part X?

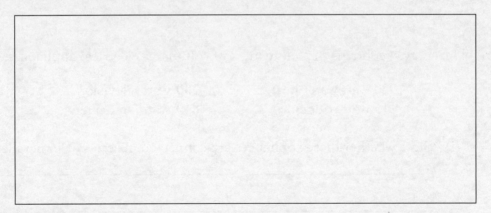

1

[Turn over

Marks

9. A caravan is connected to a mains electrical socket at a campsite.

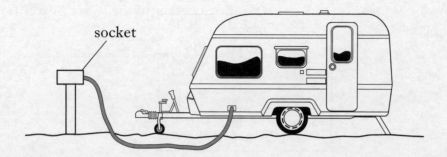

socket

(*a*) The campsite socket is fitted with a circuit breaker which has the following label attached.

> 230 volts   50 hertz a.c.
>
> Maximum current 10 amperes
>
> If circuit breaker operates,
> contact campsite reception.

What is the purpose of the circuit breaker?

2

(*b*) The caravan contains the following 230 volt mains electrical appliances.

920 watt water heater	230 watt television
115 watt fridge	1840 watt fan heater

(i) State whether these appliances are connected in series or in parallel.

Explain your answer.

2

Marks

9. **(*b*)** **(continued)**

(ii) Show by calculation that the operating current in the water heater is 4 amperes.

2

(iii) Calculate the resistance of the water heater.

2

(iv) While the water heater is operating, the fan heater is also switched on. What will happen to the circuit breaker? Use a calculation to justify your answer.

3

[Turn over

Marks

10. A student aims a ray of light at a mirror as shown. The angle of incidence of the light is 30°.

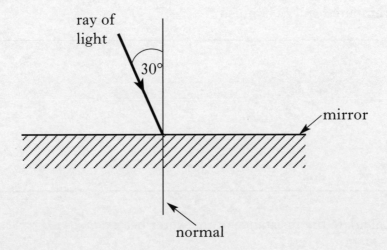

(a) On the diagram above, draw the reflected light ray. Mark the size of the angle of reflection on the diagram.

2

(b) The student now alters the angle of the ray so that it strikes the mirror along the normal.

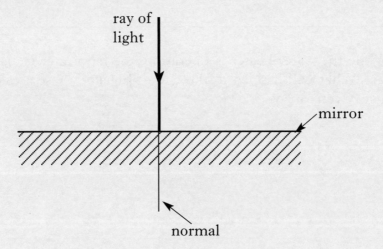

Describe below (or show on the diagram above) the new direction of the reflected ray.

1

Marks

10. (continued)

(c) (i) State **one** difference between the light produced by a laser and light from an ordinary lamp.

1

(ii) Lasers have many uses. They can be found in hospitals, factories, shops and homes.

Describe **one** use of a laser. You may include a diagram, if you wish.

2

[Turn over

Marks

11. (*a*) A student arranges a raybox to produce 3 parallel rays of light.

A lens placed in the path of the rays causes them to diverge as shown.

Complete the diagram to show the shape of this lens.

1

(*b*) The student now replaces the first lens with another one. It causes the rays to converge as shown below.

Complete the diagram to show the shape of this lens.

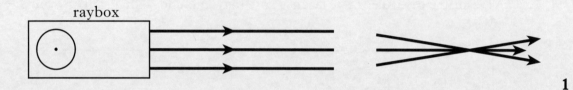

1

(*c*) This lens is replaced by a third lens. It has the following effect on the rays.

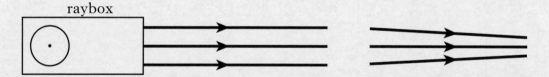

In what way is the shape of this lens different from the lens in part (*b*)?

1

(*d*) What type of lens would be chosen by an optician to correct long sight?

1

Marks

12. (*a*) The following list shows some tasks done by hospital staff.

     A   measuring the weight of a baby

     B   producing a scan of an unborn child

     C   treating a footballer's strained leg muscle

     D   treating a brain tumour without surgery

     E   treating a child's eczema

     F   detecting a break in an arm-bone

     G   measuring a patient's body temperature

From the above list, choose **one** use for each of the following by placing one of the letters A–G in the box.

Infrared ☐   X-rays ☐

Ultrasound ☐   Ultraviolet ☐   2

(*b*) Which one of these cannot pass through a vacuum? **Circle your answer**.

    Infrared   X-rays   Ultrasound   Ultraviolet   1

[Turn over

Marks

13. An engineering company has a gamma radiation source for testing welded joints.

    (a) Give **one** precaution which is taken to reduce the risk to the workers from the radiation.

    1

    (b) A safety officer measures the count rate from the gamma radiation at different distances from the source.

    The graph shows how the count rate varies with distance.

    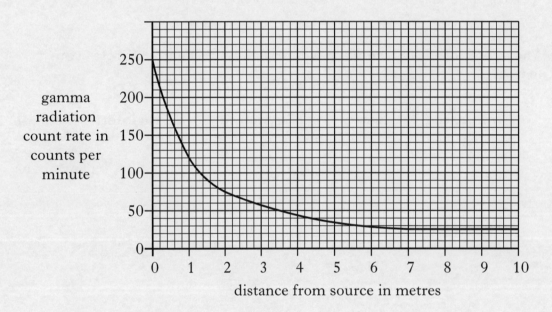

    (i) What distance from the source gives a count rate of 50 counts per minute?

    1

DO NOT
WRITE IN
THIS
MARGIN

Marks

13. **(b) (continued)**

(ii) Explain why the count rate does not drop to zero at large distances from the source.

1

[Turn over

Marks

14. (*a*) The diagrams show two sound level meter readings taken in a factory. One was taken close to a drilling machine. The other was taken in the office.

   (i) In the box below each diagram, write where the measurement was taken.

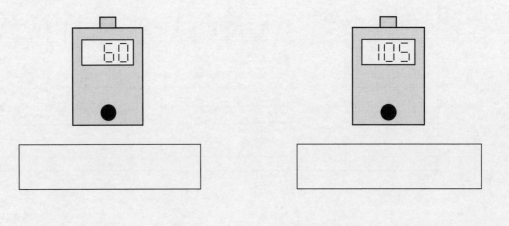

1

   (ii) What is the unit of sound level?

1

   (iii) Explain why ear protection should be worn by people working near the drilling machine.

2

(*b*) The factory technician tests different types of ear protector. A dummy head is fitted with a small microphone inside one ear. The microphone is connected to an oscilloscope.

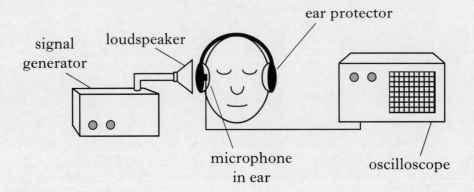

signal generator   loudspeaker   ear protector   oscilloscope   microphone in ear

Three different models of ear protector are tested. Each model is tested at high and low frequencies. For each model, the two oscilloscope traces are shown on the next page.

Marks

14. **(b)** **(continued)**

Model X	Model Y	Model Z

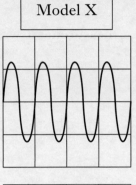 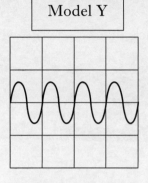 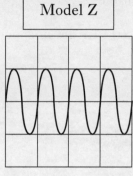

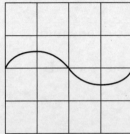 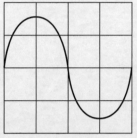 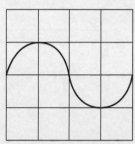

(i) Which model of ear protector gives best protection against high frequency sound? You **must** explain your answer.

2

(ii) Which model of ear protector gives best protection against low frequency sound? You **must** explain your answer.

2

[Turn over

Marks

15. A high speed train takes its energy from overhead wires, supported on masts as shown.

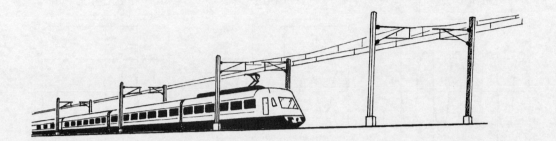

(*a*) On one section of the journey, the train passes 180 masts in one minute.

   (i) Calculate the number of masts the train passes in one second.

1

   (ii) The distances between masts are each 32 metres. Calculate the average speed of the train on this section of its journey.

2

DO NOT WRITE IN THIS MARGIN

Marks

15. (continued)

(*b*) The train wheels are driven by electric motors. At maximum speed, the total power requirement is 12 000 000 watts.

The voltage of the overhead wires is 25 000 volts.

Calculate the current taken from the overhead wires when the train travels at maximum speed.

2

[Turn over

Marks

16. A scientist tests safety helmets for cyclists. A helmet is fitted to a dummy head and dropped on to a concrete surface. A new helmet of the same type is used for each test.

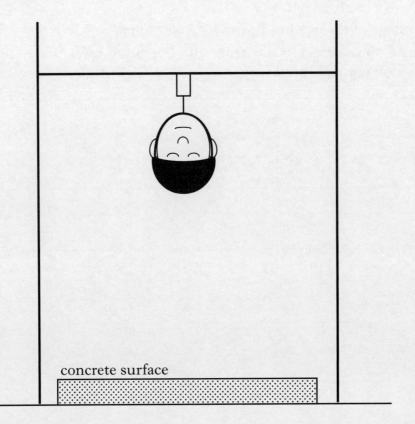

concrete surface

(a) In one test, a dummy head of mass 4·8 kilograms is fitted with a helmet of mass 0·2 kilograms.

(i) Calculate the mass of the combined head and helmet.

1

(ii) Calculate the weight of the combined head and helmet.

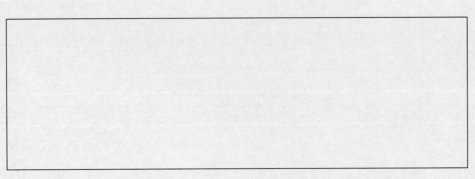

2

Marks

16. (*a*) **(continued)**

(iii) Explain what is meant by the weight of an object.

1

(iv) Which instrument would you use to measure weight?

1

(*b*) Describe and explain the effect on the damage to a helmet of each of the following changes.

(i) Exchange the dummy head for one of mass 6·0 kilograms.

2

(ii) Increase the height of the drop.

2

[Turn over

Marks

17. The diagram shows the electronic system which controls a car's air conditioning unit.

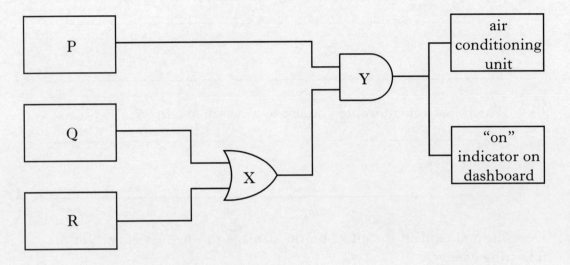

(a) (i) What type of logic gate is X?

1

(ii) What type of logic gate is Y?

1

(b) The air conditioning unit cools the inside of the car if it is too hot while the ignition switch is on. It also works if it is too hot while the ignition switch is off, if a separate manual switch is switched on instead.

The system includes a temperature sensor. An indicator on the dashboard lights up when the air conditioning unit is working.

Use the letters P, Q and R from the diagram above to complete the table showing the positions of the input devices in the electronic system.

Device	Logic levels		Position
Ignition switch	off **0**	on **1**	
Manual switch	off **0**	on **1**	
Temperature sensor	cold **0**	hot **1**	

3

DO NOT WRITE IN THIS MARGIN

Marks

17. (continued)

(*c*) A list of electronic components is shown below.

electric motor	**LDR**	**microphone**	**switch**
loudspeaker	**thermistor**	**lamp**	**LED**

(i) Which component **from the list** could be used in the temperature sensor? **1**

(ii) Which component **from the list** could be used for the air conditioning "on" indicator? **1**

(iii) The air conditioner includes an electric motor.
What is the energy change in an electric motor? **1**

[Turn over

Marks

18. An engineer in a hi-fi shop demonstrates a sound system to a customer. The diagram shows the connections between the subsystems.

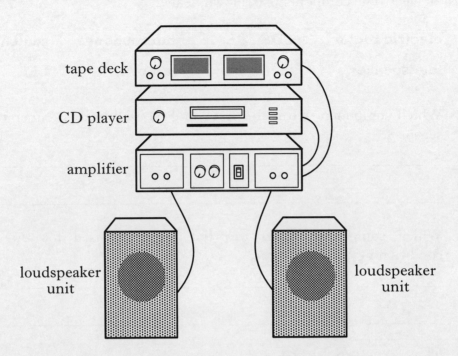

(a) From the diagram, name

   (i) an input subsystem

   1

   (ii) an output subsystem.

   1

(b) The customer asks the engineer to measure the voltage gain of the amplifier. The engineer connects a signal generator and two voltmeters to the amplifier as shown below.

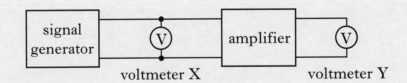

The readings on the two voltmeters are 18·0 volts and 0·002 volts.

Marks

18. (b) (continued)

(i) Which of these readings is taken from voltmeter Y?

1

(ii) Calculate the voltage gain of this amplifier.

2

[END OF QUESTION PAPER]

Page twenty-seven

Marks

**YOU MAY USE THE SPACE ON THIS PAGE TO REWRITE ANY ANSWER
YOU HAVE DECIDED TO CHANGE IN THE MAIN PART OF THE ANSWER
BOOKLET. TAKE CARE TO WRITE IN CAREFULLY THE APPROPRIATE
QUESTION NUMBER.**

[BLANK PAGE]

FOR OFFICIAL USE

Total Marks

X069/101

NATIONAL
QUALIFICATIONS
2006

WEDNESDAY, 17 MAY
1.00 PM – 2.30 PM

PHYSICS
INTERMEDIATE 1

Fill in these boxes and read what is printed below.

Full name of centre

Town

Forename(s)

Surname

Date of birth
Day Month Year

Scottish candidate number

Number of seat

Reference may be made to the Physics Data Booklet.

1 All questions should be answered.

2 The questions may be answered in any order but all answers must be written clearly and legibly in this book.

3 For each of questions 1–7 there is only **one** correct answer. Write down, in the space provided, the letter corresponding to the answer you think is correct.

4 For questions 8–21 write your answer where indicated by the question or in the space provided at the end of the answer book.

5 If you change your mind about your answer you may score it out and rewrite it in the space provided at the end of the answer book.

6 Before leaving the examination room you must give this book to the invigilator. If you do not, you may lose all the marks for this paper.

SCOTTISH
QUALIFICATIONS
AUTHORITY

Marks

1. Which colours of light are mixed on a colour TV screen?

   A   Red, blue and yellow
   B   Red, green and yellow
   C   Blue, green and yellow
   D   Blue, green and white
   E   Blue, green and red

   Answer [   ]   1

2. Which row in the table describes the components in a telephone handset?

	Mouthpiece	Earpiece
A	loudspeaker	microphone
B	amplifier	microphone
C	microphone	transmitter
D	microphone	loudspeaker
E	receiver	loudspeaker

   Answer [   ]   1

3. Which row in the table describes an eye defect?

	Defect	Near object	Distant object
A	long sight	blurred	clear
B	long sight	clear	clear
C	long sight	clear	blurred
D	short sight	blurred	clear
E	short sight	blurred	blurred

   Answer [   ]   1

4. Sound level is measured in

   A   hertz
   B   joules
   C   decibels
   D   hertz per second
   E   metres per second.

   Answer [   ]   1

Marks

5. A student makes the following three statements about a tennis ball.

   I If the ball is hit with more force it will have a greater speed.

  II The surface of the court does not affect the height of the bounce.

III The angle of the shot affects the range of the ball.

Which of these statements is/are correct?

A I only
B I and II only
C I and III only
D II and III only
E I, II and III

Answer ☐   **1**

6. Which of the following block diagrams shows the main parts of an electronic system?

A   **Process** → **Input** → **Output**

B   **Process** → **Output** → **Input**

C   **Output** → **Process** → **Input**

D   **Input** → **Process** → **Output**

E   **Input** → **Output** → **Process**

Answer ☐   **1**

7. A thermistor is

A an output device with a constant resistance
B an output device whose resistance changes with light level
C an output device whose resistance changes with temperature
D an input device whose resistance changes with light level
E an input device whose resistance changes with temperature.

Answer ☐   **1**

[Turn over

DO NOT
WRITE IN
THIS
MARGIN

Marks

8. Read the following passage taken from a student's diary.

"Gran thinks my new picture mobile phone is like magic. Wait till she sees Dad's hand held TV! She can remember when her family got their first wireless. There were only a few radio stations at that time. Gran now likes to listen to news programmes and Scottish music. I prefer to listen to Rock 104 FM."

(*a*) Why is a radio sometimes called a *wireless*?

1

(*b*) In the sentence below, circle **one** word in the box to make the statement correct.

A radio signal has a | higher/lower | frequency than a TV signal.　　1

(*c*) State the speed of radio and TV waves in air.

1

Marks

8. (continued)

(*d*) An engineer has to choose suitable frequencies for radio stations W, X, Y and Z. The transmissions from the different stations must not interfere with each other. Only three frequencies are available for the four stations. Complete the table to show the frequency for each transmitter.

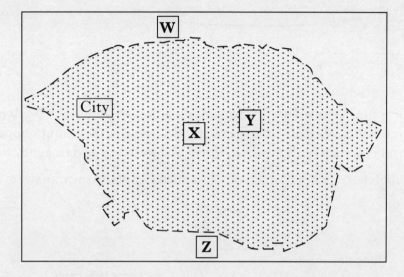

Transmitter	*Frequency* in millions of hertz
	89·4
	94·3
	94·3
	98·2

1

[Turn over

Marks

9. (*a*) Complete the path of the ray of light through the optical fibre shown below.

1

(*b*) Two computer networks are connected to the same aerial. Network A is connected using optical fibre, network B is connected using copper wire. The aerial is struck by lightning. Only one network is damaged.

State which network is damaged. You **must** explain your answer.

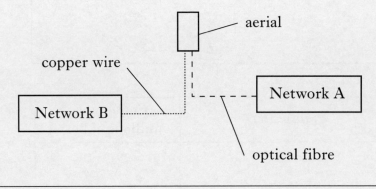

2

(*c*) Give **one** disadvantage of using optical fibre in a communication system.

1

Marks

10. A student uses a pair of ceramic hair straighteners. They operate at 230 volts with a power rating of 805 watts.

(a) (i) Calculate the current in the straighteners when switched on.

2

(ii) Which fuse should be used in the plug attached to the straighteners? The following fuses are available. **Circle your choice**.

3 amperes   5 amperes   13 amperes

1

(b) The straighteners stop working. The student removes the fuse from the plug to test it. The student makes a continuity tester using a battery, lamp and wires.

(i) Draw a circuit diagram to show how the continuity tester is used to test the fuse.

2

(ii) The lamp does not light. The student concludes that the fuse has blown. What name is given to this fault?

1

(c) The ceramic straighteners give off heat radiation. Give another name for heat radiation.

1

Marks

11. The following circuit diagram shows a fan heater which can provide cold, warm or hot air.

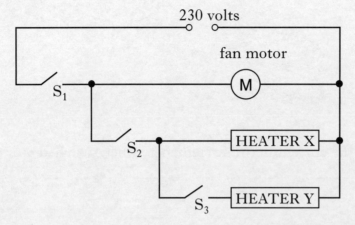

230 volts

fan motor

M

HEATER X

HEATER Y

S_1

S_2

S_3

(a) (i) Which switch or switches must be closed so that the heater provides a stream of warm air?

1

(ii) State the energy change that takes place in the motor.

1

(b) The supply voltage is 230 volts.

(i) State the voltage across the motor when the fan is on.

1

(ii) State the voltage across heater X when the heater is set at hot.

1

Marks

12. A student builds the following circuit.

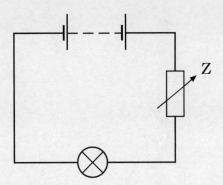

(a) What is component Z?

1

(b) The current in the lamp is 0·5 amperes. The voltage across the lamp is 4 volts. Calculate the resistance of the lamp.

2

(c) Suggest **two** changes you could make to the circuit to make the lamp brighter.

2

[Turn over

Marks

13. A spectator at the Open Golf Championship uses a viewer to see over the heads of the crowd.

(a) What happens to a ray of light when it hits a mirror?

1

(b) Complete the diagram below to show the path of the ray of light through the viewer to the spectator's eye.

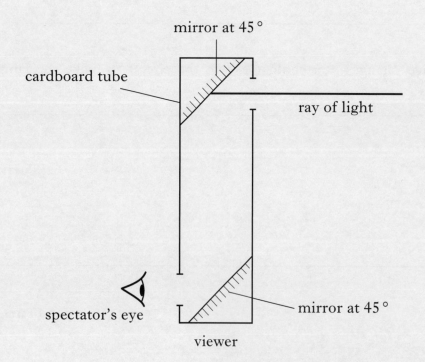

2

Marks

13. (continued)

(*c*) The spectator uses his mobile phone to tell a friend the result of the game.

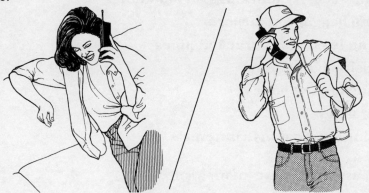

Complete the sentences below by choosing from the wordbank. Words may be used more than once.

sound	**microphone**	**radio**	**loudspeaker**
LED	**electrical**	**LDR**	**light**

The mouthpiece of the mobile phone contains a ☐ .

This device changes ☐ energy to ☐ energy.

An ☐ is used to light up the keypad.

This device changes ☐ energy to ☐ energy.　**3**

[Turn over

Marks

14. Shown below are some uses of different types of radiation.

> **treating muscle strain**
> **checking security markings on banknotes**
> **scanning baggage at airports**
> **detecting leaks from cracked pipes**
> **night vision binoculars**
> **detecting broken bones**
> **treating acne**
> **sterilising surgical instruments**

Complete the table using the above information.

Only **one** entry should appear in each box.

Radiation	Medical uses	Non-medical uses
X-rays		
Gamma		
Infrared		
Ultraviolet		

4

Page twelve

Marks

15. A doctor uses the device shown to examine the inside of a patient's windpipe.

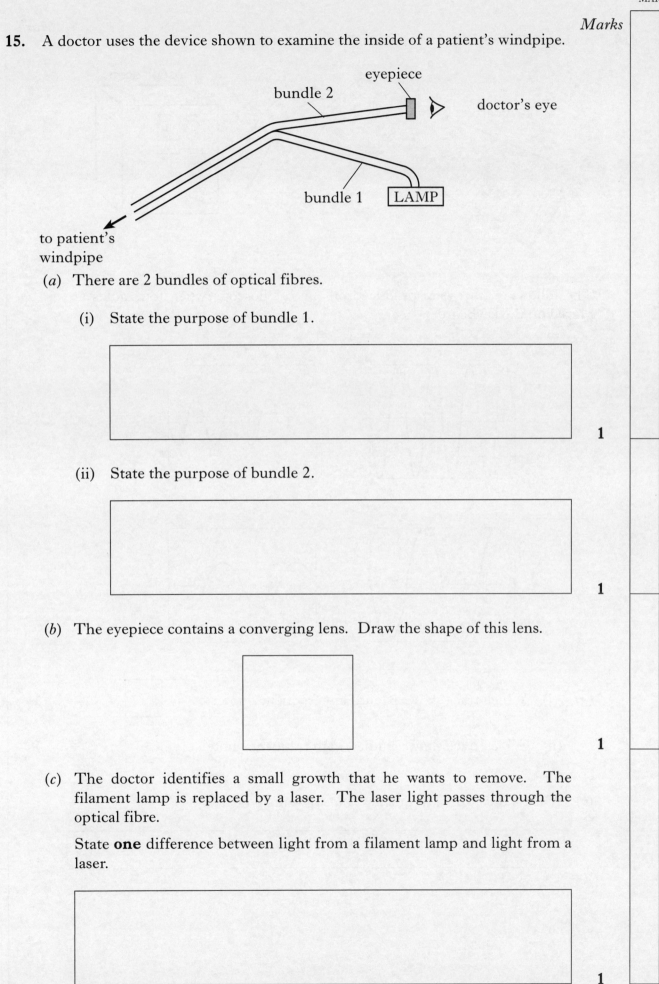

(*a*) There are 2 bundles of optical fibres.

    (i) State the purpose of bundle 1.

1

    (ii) State the purpose of bundle 2.

1

(*b*) The eyepiece contains a converging lens. Draw the shape of this lens.

1

(*c*) The doctor identifies a small growth that he wants to remove. The filament lamp is replaced by a laser. The laser light passes through the optical fibre.

State **one** difference between light from a filament lamp and light from a laser.

1

Marks

16. An electronic keyboard is connected to an oscilloscope.

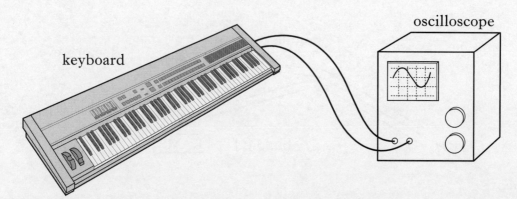

The following traces are produced on the oscilloscope when four notes are played on the keyboard.

P

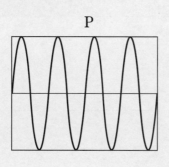

Q

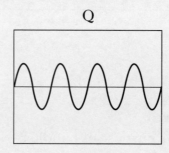

R

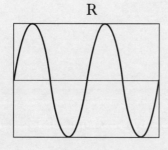

S

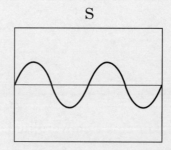

(a) (i) Which trace shows a loud, low frequency note?

1

(ii) Which trace shows a quiet, high frequency note?

1

(b) What is the normal frequency range of human hearing?

1

Marks

16. **(continued)**

(*c*) A sound sensor is placed 1·3 metres from the keyboard's loudspeaker.

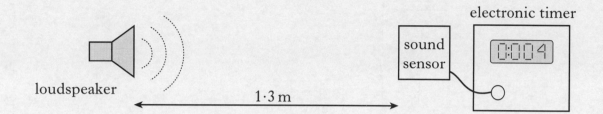

electronic timer

sound
sensor

0:004

loudspeaker

1·3 m

An electronic timer records that the sound takes 0·004 seconds to travel from the loudspeaker to the sensor. Use these figures to calculate the speed of the sound in air.

2

(*d*) A student records her friend singing. They listen to the recording.

tape recorder

The singer does not recognise her recorded voice but her friend says that it sounds just like her. Explain why.

2

Marks

17. A girl dives from the 10 metre board. She has a mass of 65 kilograms.

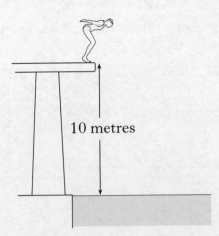

10 metres

(*a*) Calculate the weight of the diver.

2

(*b*) The following graph shows the downward speed of the diver.

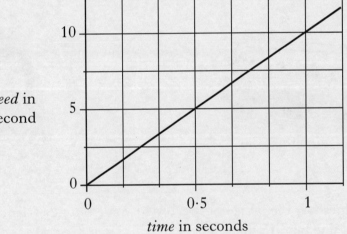

downward speed in
metres per second

time in seconds

    (i)   What is the speed of the diver after 0·5 seconds?

1

    (ii) What is the speed of the diver after 1 second?

1

    (iii) What name is given to this increase in speed?

1

DO NOT WRITE IN THIS MARGIN

Marks

17. (continued)

(*c*) The diver takes one second to fall the first 5 metres. Will she take more, less or the same time to fall the second 5 metres? You **must** give a reason for your answer.

2

(*d*) The diver enters the water.

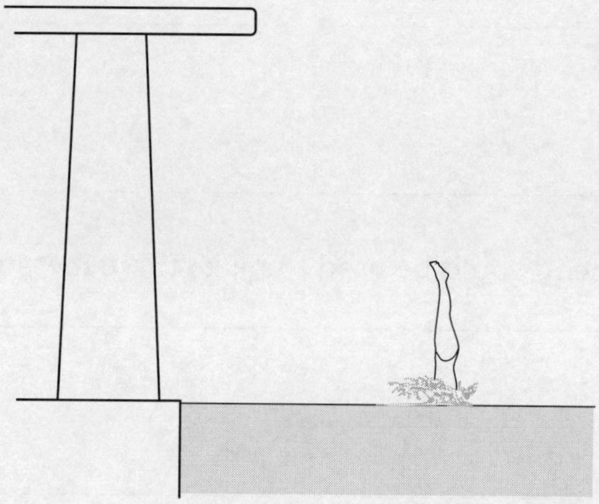

Explain why her speed decreases on entering the water.

1

[Turn over

Marks

18. Two music students are playing the cymbals in the school band. One of the students says that the cymbals can be used in an experiment to measure the speed of sound in the school grounds.

(a) (i) What measurements would be made to measure the speed of sound?

1

(ii) What equipment would be used to make these measurements?

1

(iii) How would these measurements be used to calculate the speed of the sound?

1

(b) Suggest **one** factor which could make the students' measurements inaccurate.

1

Marks

19. A car is driven along a road. The two horizontal forces acting on the car are the engine force and friction.

(*a*) What can you say about the sizes and directions of these forces when the car is travelling at a steady speed?

2

(*b*) The car travels a distance of 8100 metres through a city. The journey takes 15 minutes.

Calculate the average speed of the car in metres per second.

2

(*c*) The average speed for this journey is less than the maximum speed. Explain why.

1

[Turn over

Marks

20. (*a*) A list of electronic devices is shown.

motor	**switch**	**AND gate**
LED	**buzzer**	**LDR**

(i) From the list select one process device.

1

(ii) From the list select one output device.

1

(iii) From the list select one input device.

1

(iv) Draw the symbol for the AND gate.

1

(*b*) A public address system consists of a microphone, amplifier and loudspeaker. The voltage across the microphone is 0·05 volts. The voltage across the loudspeaker is 20 volts. Calculate the voltage gain of the amplifier.

2

Marks

21. A student investigates the logic gates shown.

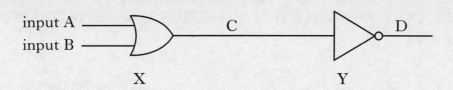

(*a*) (i) Name logic gate X. 1

(ii) Name logic gate Y. 1

(*b*) (i) Is a high input voltage known as logic 0 or logic 1? 1

(ii) The diagrams below show oscilloscope traces for input A and input B.

Draw the oscilloscope traces for output C and output D.

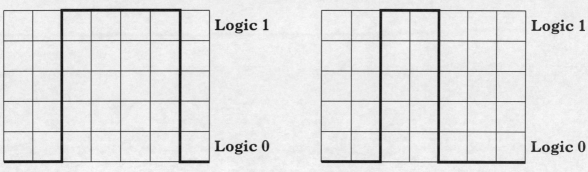

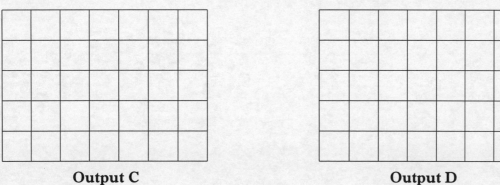

2

[END OF QUESTION PAPER]

Marks

YOU MAY USE THE SPACE ON THIS PAGE TO REWRITE ANY ANSWER YOU HAVE DECIDED TO CHANGE IN THE MAIN PART OF THE ANSWER BOOKLET. TAKE CARE TO WRITE IN CAREFULLY THE APPROPRIATE QUESTION NUMBER.

[BLANK PAGE]

FOR OFFICIAL USE

Total Marks

X069/101

NATIONAL
QUALIFICATIONS
2007

WEDNESDAY, 16 MAY
1.00 PM – 2.30 PM

PHYSICS
INTERMEDIATE 1

Fill in these boxes and read what is printed below.

Full name of centre

Town

Forename(s)

Surname

Date of birth
Day Month Year

Scottish candidate number

Number of seat

Reference may be made to the Physics Data Booklet.

Section A – Questions 1–20 (20 marks)

Instructions for completion of **Section A** are given on page two.

For this section of the examination you must use an **HB pencil**.

Section B (60 marks)

All questions should be attempted.

The questions may be answered in any order but all answers are to be written in the spaces provided in this answer book, **and must be written clearly and legibly in ink**.

Rough work, if any should be necessary, should be written in this book, and then scored through when the fair copy has been written. If further space is required, a supplementary sheet for rough work may be obtained from the invigilator.

Additional space for answers will be found at the end of the book. If further space is required, supplementary sheets may be obtained from the invigilator and should be inserted inside the **front** cover of this booklet.

Before leaving the examination room you must give this book to the invigilator. If you do not, you may lose all the marks for this paper.

SCOTTISH
QUALIFICATIONS
AUTHORITY

SECTION A

1 Check that the answer sheet provided is for Physics Intermediate 1 (Section A).

2 For this section of the examination you must use an **HB pencil** and, where necessary, an eraser.

3 Check that the answer sheet you have been given has **your name**, **date of birth**, **SCN** (Scottish Candidate Number) and **Centre Name** printed on it.

Do not change any of these details.

4 If any of this information is wrong, tell the Invigilator immediately.

5 If this information is correct, **print** your name and seat number in the boxes provided.

6 The answer to each question is **either** A, B, C, D or E. Decide what your answer is, then, using your pencil, put a horizontal line in the space provided (see sample question below).

7 There is **only one correct** answer to each question.

8 Any rough working should be done on the question paper or the rough working sheet, **not** on your answer sheet.

9 At the end of the exam, put the **answer sheet for Section A inside the front cover of this answer book**.

Sample Question

The energy unit measured by the electricity meter in your home is the

    A kilowatt-hour

    B ampere

    C watt

    D coulomb

    E volt.

The correct answer is **A**—kilowatt-hour. The answer **A** has been clearly marked in **pencil** with a horizontal line (see below).

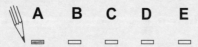

Changing an answer

If you decide to change your answer, carefully erase your first answer and, using your pencil, fill in the answer you want. The answer below has been changed to **E**.

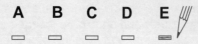

SECTION A

Answer questions 1–20 on the answer sheet.

1. On a colour television screen, which colours of light are mixed to produce magenta?

   A   Red and green

   B   Red and blue

   C   Blue and green

   D   Blue and yellow

   E   Blue, green and red

2. The unit of frequency is the

   A   decibel

   B   watt

   C   hertz

   D   volt

   E   second.

3. A student makes the following statements about optical fibres.

   I   Light travels in an optical fibre at a speed of nearly 200 000 000 metres per second.

   II   Optical fibres can be made from long, thin pieces of glass.

   III   Optical fibres carry electrical signals.

   Which of these statements is/are true?

   A   I only

   B   I and II only

   C   I and III only

   D   II and III only

   E   I, II and III

[Turn over

4. Which row in the table shows the symbols for a resistor and a variable resistor?

	Resistor	Variable Resistor
A		
B		
C		
D		
E		

5. A circuit is set up as shown.

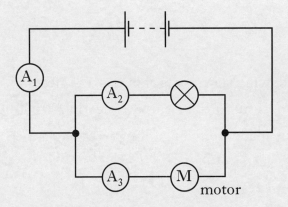

Which row in the table shows the possible readings on ammeters A_1, A_2 and A_3?

	Reading on A_1 in amperes	Reading on A_2 in amperes	Reading on A_3 in amperes
A	1	2	3
B	1	3	2
C	3	3	3
D	5	2	3
E	6	2	3

6. Infrared radiation is used in medicine to

   A   treat muscle injuries

   B   detect broken bones

   C   trace blood flow

   D   treat acne

   E   kill cancer cells.

7. Which of the following can be used to detect X-rays?

   A   Photographic film

   B   Aerial

   C   Microphone

   D   Oscilloscope

   E   Continuity tester

8. Which of the following best describes a laser beam?

   A   A narrow beam of white light

   B   A wide beam of white light

   C   A wide beam of red light

   D   A narrow beam of green light

   E   A wide beam of green light

9. Dogs can hear sounds with frequencies between 10 hertz and 30 000 hertz.
Dogs and humans can both hear sounds with a frequency of

   A   10 hertz

   B   15 hertz

   C   1500 hertz

   D 25 000 hertz

   E 30 000 hertz.

10. A student carries out an experiment with a vibrating guitar string.
Which of the following changes would produce the lowest frequency of sound?

   A   Halving the length of the string

   B   Doubling the length of the string

   C   Keeping the length the same and tightening the string

   D   Halving the length of the string and tightening the string

   E   Doubling the length of the string and tightening the string

[Turn over

11. A student is 1500 metres from a fireworks display.

He hears the bang from an exploding firework 5 seconds after seeing the flash.

Using these figures, the student calculates the speed of sound in air to be

A   5 metres per second

B   300 metres per second

C   340 metres per second

D   1500 metres per second

E   7500 metres per second.

12. An alarm clock is inside a jar as shown.

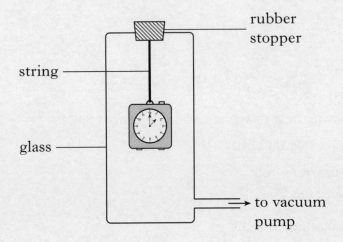

The sound from the alarm is heard clearly.

All of the air in the jar is now pumped out.

The alarm cannot be heard because

A   sound cannot pass through string

B   sound cannot pass through a vacuum

C   sound cannot pass through glass

D   sound cannot pass through air

E   sound cannot pass through rubber.

13. A CD player contains an amplifier.

The amplifier is used to

A   increase the frequency of an electrical signal

B   decrease the frequency of an electrical signal

C   change sound into electrical energy

D   change electrical energy into sound

E   increase the amplitude of an electrical signal.

14. Five cars are acted on by different forces.

Which of the cars is speeding up to the right?

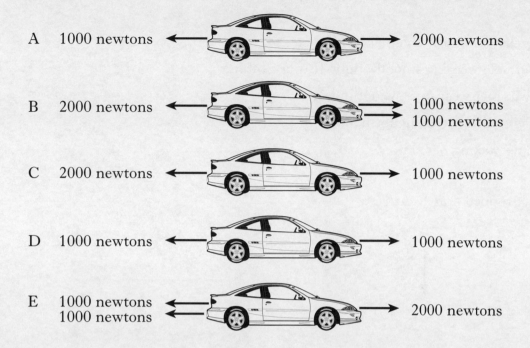

A   1000 newtons ← → 2000 newtons

B   2000 newtons ← → 1000 newtons
→ 1000 newtons

C   2000 newtons ← → 1000 newtons

D   1000 newtons ← → 1000 newtons

E   1000 newtons ←
1000 newtons ← → 2000 newtons

15. A car moves along a straight road in the direction shown.

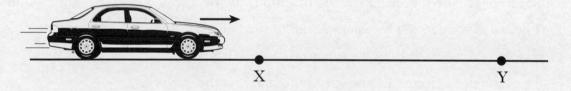

X   Y

The distance between X and Y is measured.

The only other measurement needed to calculate the average speed of the car between X and Y is

A   the speed of the car at X

B   the speed of the car at Y

C   the acceleration of the car between X and Y

D   the time for the car to travel from X to Y

E   the length of the car.

[Turn over

16. Sportsmen and sportswomen use a variety of methods to increase or decrease the force of friction.

    Which of the following statements describes a method used to increase the force of friction?

    A A cyclist oiling the chain of her bicycle

    B An athlete wearing smooth, tight-fitting clothing

    C A swimmer wearing a swimming cap

    D A weightlifter chalking his hands

    E A skier waxing her skis

17. A ball is dropped onto a surface.

    A student makes the following statements about the height to which the ball rebounds.

    I The height is affected by the material of the ball.

    II The height is affected by the material of the surface.

    III The height is affected by the speed the ball hits the surface.

    Which of these statements is/are true?

    A III only

    B I and II only

    C I and III only

    D II and III only

    E I, II and III

18. Which one of the following devices is used to convert electrical energy into other forms of energy?

    A Switch

    B LDR

    C Microphone

    D Motor

    E Thermistor

19. The block diagram for a security light system is shown.

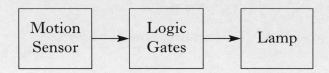

Which row in the table shows the input, process and output for this system?

	Input	Process	Output
A	Logic gates	Motion sensor	Lamp
B	Logic gates	Lamp	Motion sensor
C	Lamp	Logic gates	Motion sensor
D	Motion sensor	Lamp	Logic gates
E	Motion sensor	Logic gates	Lamp

20. Input and output devices are used in electronics.

Which row in the table is correct?

	Input device	Output device
A	LED	buzzer
B	microphone	LDR
C	loudspeaker	microphone
D	switch	microphone
E	LDR	motor

Candidates are reminded that the answer sheet for Section A MUST be placed INSIDE the front cover of this answer book.

[Turn over for Section B

DO NOT
WRITE IN
THIS
MARGIN

Marks

SECTION B

Answer questions 21–31 in the spaces provided.

21. Mobile phones are used by many people today.

    (*a*) From the list below, circle the type of signal used by mobile phones to send and receive phone calls.

    Gamma   **X-Rays**   **Ultraviolet**

    Visible Light   **Infrared**   **Radio & TV**   1

    (*b*) Name the device used in a mobile phone to change sound energy into electrical energy.

    1

    (*c*) (i) Name the device used in a mobile phone that produces sound.

    1

    (ii) What must this device do to produce sound?

    1

21. **(continued)**

Marks

(*d*) In an experiment to investigate mobile phone signals, a student sets up the following equipment. The mobile phone transmits signals as pulses.

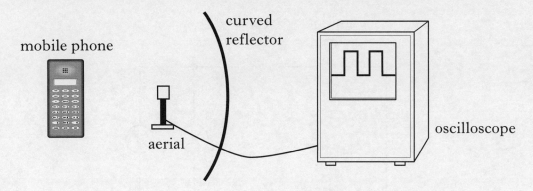

(i) Why is there a curved reflector behind the aerial?

1

(ii) Complete the diagram below to show how the curved reflector affects the signals. On your diagram, mark where the aerial should be positioned to get the strongest signal.

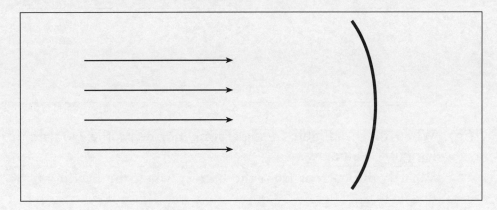

2

(iii) On the blank oscilloscope grid below, draw the signal that you would expect to see if the curved reflector was removed.

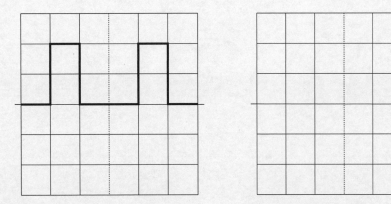

Signal received when a curved reflector is used.

Signal received with curved reflector removed.

2

Marks

22. Cars have many different electric circuits.

(*a*) This circuit shows how the headlights are connected.

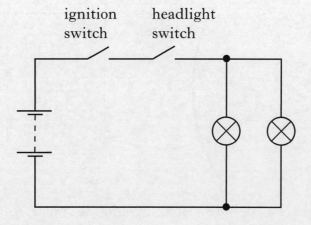

(i) Are the switches connected in series or parallel?

1

(ii) Give a reason why the switches are connected in this way.

1

(iii) When the headlights are operating normally, the current in each lamp is 4·5 amperes.

What is the current from the battery when the headlights operate normally?

1

Marks

22. (continued)

(*b*) This circuit diagram shows how the windscreen wiper motor is connected.

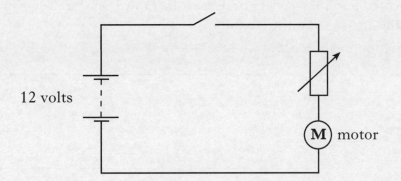

12 volts

(i) When the switch is closed, the voltage across the variable resistor is 2·4 volts. What is the voltage across the motor?

1

(ii) The current in the variable resistor is 0·6 amperes.
Calculate the resistance of the variable resistor.

2

(iii) The resistance of the variable resistor is increased.

(A) What will happen to the speed of the motor?

1

(B) Explain your answer.

1

Marks

23. A football match is broadcast live on television.

Some viewers are watching the match in a house 1 kilometre away from the stadium.

(*a*) A goal is scored and the viewers first hear the crowd on the television and then hear the crowd from the stadium.

Explain why there is a delay.

1

(*b*) Complete the block diagram below showing the main parts of a television.

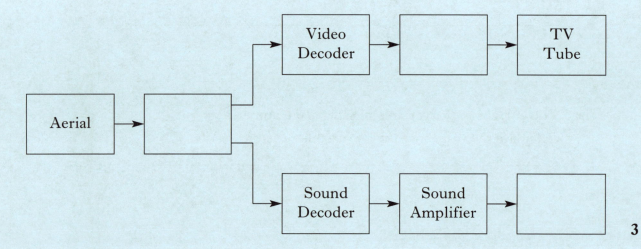

3

Page fourteen

Marks

24. A student made a telescope to look at distant objects.

The telescope has 2 lenses called the objective and eyepiece.

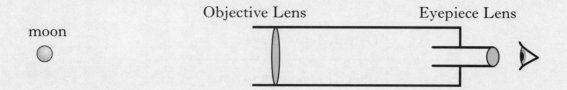

The table below gives information about the lenses.

Name	Diameter	Thickness	Purpose
Objective	Large	Thin	Collect light
Eyepiece	Small	Thick	Magnify

(a) Looking at the diagram, name the **type of lens** used as the objective.

1

(b) The student now looks through the telescope from the other end.

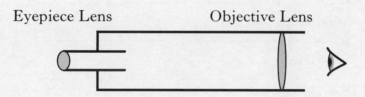

(i) Would the image now be **brighter**, **less bright** or the **same brightness** than the image he first saw in part (a)?

Explain your answer.

2

(ii) Would the image be **larger**, **smaller** or the **same size** than the image he first saw in part (a)?

Explain your answer.

2

Marks

25. Table 1 lists some uses of different types of radiation.

Table 1

	Use
A	Detected by the eye
B	Check luggage at airports
C	Control model cars
D	Grill food
E	Kills cancer cells
F	Gives a suntan

Complete Table 2 using the letters A, B, C, D, E and F, to match each radiation with its use.

You should use each letter **once** only.

Table 2

Radiation	
Gamma	
X-Rays	
Ultra-Violet	
Visible Light	
Infra Red	
Radio waves	

3

Marks

26. The diagram shows rays of light entering an eye of someone with **normal vision**.

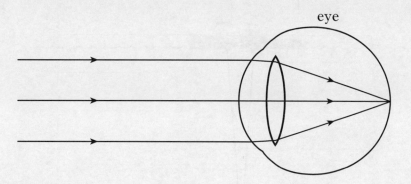

(*a*) (i) Complete the diagram below to show how rays of light from a **distant object** enter the eye of someone with **short sight**.

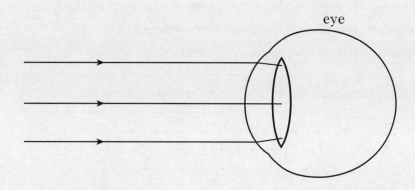

1

(ii) Complete the diagram below to show how rays of light from a **near object** enter the eye of someone with **short sight**.

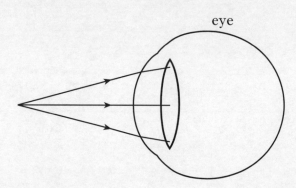

1

(*b*) Name the type of lens used to correct short sight.

1

[Turn over

Marks

27. Some boats use sound waves to find the depth of the sea.

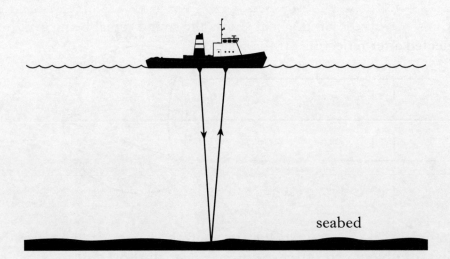

seabed

(*a*) Sound waves are sent out by a device on the bottom of the boat.
They are detected after reflecting off the seabed.
In the passage below, circle the correct words.

Sound waves are sent out by a | transmitter/receiver |

and picked up by a | transmitter/receiver | .   **1**

Marks

27. (continued)

(*b*) An oscilloscope screen on the boat shows the sound signal being sent out and detected after reflecting off the seabed.

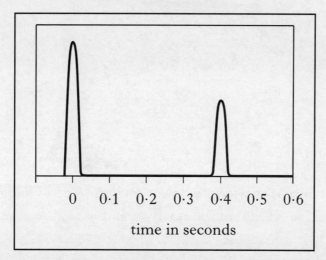

time in seconds

(i) Explain why the reflected signal has a smaller height than the signal sent out.

1

(ii) Look at the oscilloscope trace.

What is the time taken for the sound signal to reach the seabed?

1

(iii) The depth of the sea is 306 metres. Use your answer to part (ii) to calculate the speed of sound in the water.

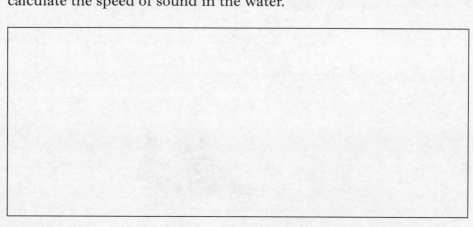

2

[Turn over

Marks

28. A skydiver jumps out of a plane.

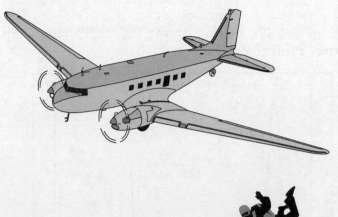

(a) The skydiver and her parachute have a total mass of 75 kilograms.
Calculate the total weight of the skydiver and parachute.

2

(b) When she first leaves the plane, the skydiver accelerates towards the earth.
What does the term accelerate mean?

1

(c) The diagram below shows the skydiver and the forces acting on her.

(i) Name the **two** vertical forces acting on the skydiver.

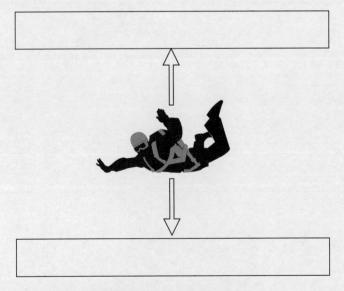

2

Marks

28. (*c*) **(continued)**

(ii) Some time later these two forces become balanced.

When the forces are balanced, does her speed **increase**, **stay the same** or **decrease**?

1

(*d*) The skydiver then opens her parachute.

(i) What happens to her speed at this moment?

1

(ii) What happens to the upward force acting on her?

1

[Turn over

Marks

29. A student playing pool knows that when the cue ball hits a cushion it obeys the same rule as light reflecting off a mirror.

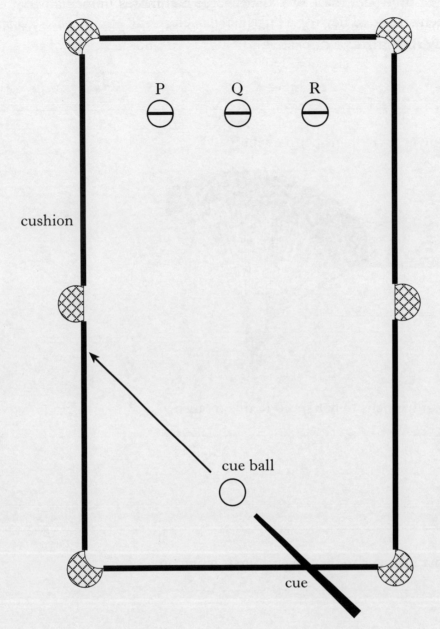

(a) Using this information, state which ball she will hit.

1

Marks

29. (continued)

(b) Her friend wants to know how fast the cue ball moves immediately after
it has been struck. They use a light gate, an electronic timer and a ruler.

Describe how this equipment would be used to find the speed of the ball
immediately after it has been struck.

Your description should include:

- the measurements taken
- how these measurements would be used to calculate the
instantaneous speed.

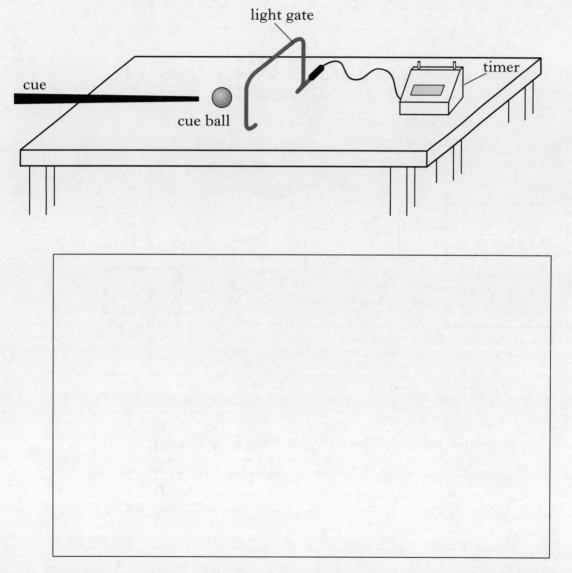

2

[Turn over

Marks

30. A museum has a statue on display. A security system is in place to protect the statue.

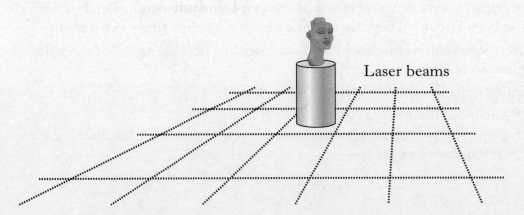

Laser beams

(a) The system uses laser beams which will sound an alarm if broken.
The diagram of the laser alarm system is shown below.
The system is activated by a master switch.

Laser beam broken	Logic 0
Laser beam not broken	Logic 1

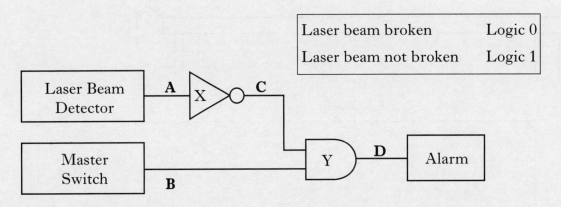

(i) Name logic gate **Y**.   1

(ii) Complete the table below to show the logic levels at **C** and **D**.

A	B	C	D
0	0		
0	1		
1	0		
1	1		

2

Marks

30. (continued)

(*b*) To improve security, the statue is placed on a pressure pad.

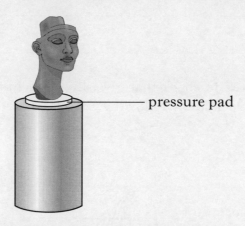

pressure pad

If the statue is lifted from the pressure pad, a small voltage is produced. This voltage has to be amplified to operate the alarm.

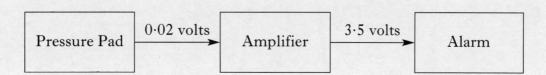

| Pressure Pad | 0·02 volts → | Amplifier | 3·5 volts → | Alarm |

(i) Calculate the voltage gain of the amplifier.

2

(ii) The power rating of the alarm is 14 W.

Calculate the current in the alarm when it operates.

2

[Turn over

Marks

31. Washing machines use thermistors as temperature sensors.

(*a*) Is a thermistor an **input**, **process** or an **output** device?

1

(*b*) A student wants to check that a washing machine thermistor is working correctly. He sets up the following experiment.

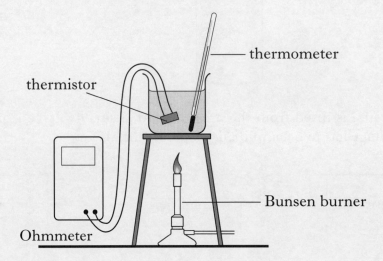

Explain how he should use the equipment to investigate how the resistance of the thermistor is affected by temperature.

Your answer should include:

- how he used the equipment
- what he measured.

2

Marks

31. **(continued)**

(c) After the experiment, the student plotted the following graph.

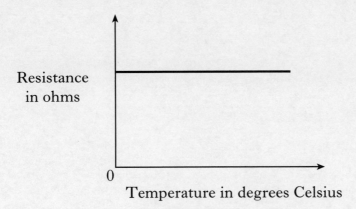

(i) Was the thermistor working correctly?

1

(ii) Explain your answer.

1

[END OF QUESTION PAPER]

Marks

YOU MAY USE THE SPACE ON THIS PAGE TO REWRITE ANY ANSWER YOU HAVE DECIDED TO CHANGE IN THE MAIN PART OF THE ANSWER BOOKLET. TAKE CARE TO WRITE IN CAREFULLY THE APPROPRIATE QUESTION NUMBER.

[BLANK PAGE]

FOR OFFICIAL USE

Total Marks

X069/101

NATIONAL   FRIDAY, 23 MAY   PHYSICS
QUALIFICATIONS 1.00 PM – 2.30 PM
2008   INTERMEDIATE 1

Fill in these boxes and read what is printed below.

Full name of centre

Town

Forename(s)

Surname

Date of birth
  Day Month Year   Scottish candidate number   Number of seat

Reference may be made to the Physics Data Booklet.

Section A – Questions 1–20 (20 marks)

Instructions for completion of **Section A** are given on page two.

For this section of the examination you must use an **HB pencil**.

Section B (60 marks)

All questions should be attempted.

The questions may be answered in any order but all answers are to be written in the spaces provided in this answer book, **and must be written clearly and legibly in ink**.

Rough work, if any should be necessary, should be written in this book, and then scored through when the fair copy has been written. If further space is required, a supplementary sheet for rough work may be obtained from the invigilator.

Additional space for answers will be found at the end of the book. If further space is required, supplementary sheets may be obtained from the invigilator and should be inserted inside the **front** cover of this booklet.

Before leaving the examination room you must give this book to the invigilator. If you do not, you may lose all the marks for this paper.

SECTION A

1 Check that the answer sheet provided is for Physics Intermediate 1 (Section A).

2 For this section of the examination you must use an **HB pencil** and, where necessary, an eraser.

3 Check that the answer sheet you have been given has **your name**, **date of birth**, **SCN** (Scottish Candidate Number) and **Centre Name** printed on it.

Do not change any of these details.

4 If any of this information is wrong, tell the Invigilator immediately.

5 If this information is correct, **print** your name and seat number in the boxes provided.

6 The answer to each question is **either** A, B, C, D or E. Decide what your answer is, then, using your pencil, put a horizontal line in the space provided (see sample question below).

7 There is **only one correct** answer to each question.

8 Any rough working should be done on the question paper or the rough working sheet, **not** on your answer sheet.

9 At the end of the exam, put the **answer sheet for Section A inside the front cover of this answer book**.

Sample Question

The energy unit measured by the electricity meter in your home is the

    A kilowatt-hour

    B ampere

    C watt

    D coulomb

    E volt.

The correct answer is **A**—kilowatt-hour. The answer **A** has been clearly marked in **pencil** with a horizontal line (see below).

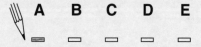

Changing an answer

If you decide to change your answer, carefully erase your first answer and, using your pencil, fill in the answer you want. The answer below has been changed to **E**.

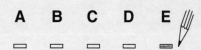

SECTION A

Answer questions 1–20 on the answer sheet.

1. Which of the following colours of light can be mixed to give all the colours seen on a television screen?

   A   Red, yellow and blue

   B   Red, green and blue

   C   Blue, magenta and red

   D   Red, orange and yellow

   E   Cyan, green and blue

2. Optical fibres are used to transmit

   A   electrical signals

   B   light signals

   C   ultrasound signals

   D   radio signals

   E   sound signals.

3. A fax machine can send information from one location to another.

   A student makes the following statements about a fax machine.

     I   A diagram can be sent by fax.

    II   Documents are faxed at the speed of sound.

   III   Faxes can be sent by telephone.

   Which of these statements is/are correct?

   A   I only

   B   I and II only

   C   I and III only

   D   II and III only

   E   I, II and III

[Turn over

4. A telephone handset contains a mouthpiece and an earpiece.

   Which row in the table shows the energy changes in the mouthpiece and the earpiece?

	Mouthpiece	Earpiece
A	sound → light	electrical → sound
B	sound → electrical	electrical → sound
C	electrical → sound	sound → electrical
D	electrical → sound	electrical → sound
E	sound → electrical	light → sound

5. A circuit is set up as shown.

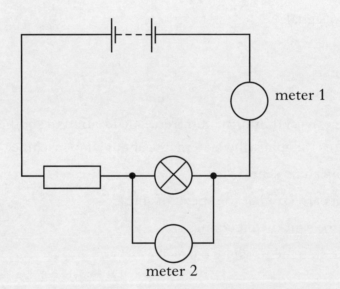

   Which row in the table shows the measurements displayed by meter 1 and meter 2?

	Meter 1	Meter 2
A	voltage across lamp	current in lamp
B	voltage across resistor	voltage across lamp
C	current in resistor	voltage across lamp
D	current in lamp	voltage across battery
E	current in resistor	current in lamp

6. A student states that telephone signals can be sent between a transmitter and a receiver in the following ways.

   I Using electrical signals in metal wires

  II Using light signals in optical fibres

III Using radio signals in air

Which of these statements is/are correct?

A I only

B II only

C I and II only

D I and III only

E I, II and III

7. The diagram shows a mains flex connected correctly to a plug.

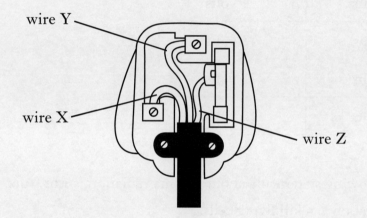

Which row in the table shows the colours of the insulation on wires X, Y and Z?

	Wire X	*Wire Y*	*Wire Z*
A	brown	blue	green/yellow
B	green/yellow	brown	blue
C	blue	brown	green/yellow
D	brown	green/yellow	blue
E	blue	green/yellow	brown

[Turn over

8. A circuit is set up as shown.

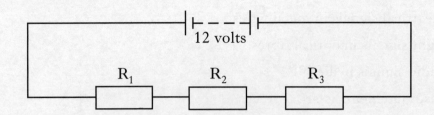

The voltage of the battery is 12 volts. The voltage across resistor R_1 is 5 volts. Which row in the table shows possible voltages across R_2 and R_3?

	Voltage across R_2	Voltage across R_3
A	2 volts	3 volts
B	3 volts	4 volts
C	5 volts	5 volts
D	5 volts	12 volts
E	12 volts	12 volts

9. Which of the following statements about gamma radiation is **not** true?

A Gamma radiation can kill living cells.

B Gamma radiation can pass through most materials.

C Gamma radiation is present in our surroundings.

D Gamma radiation can be used as a tracer.

E Gamma radiation is visible to the naked eye.

10. The diagram shows rays of light in an eye that has a defect.

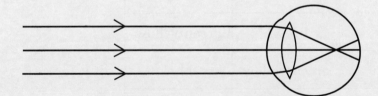

Which shape of lens could be used to correct this defect?

A   B   C   D   E

11. Which of the following graphs shows how the strength of a radioactive source changes with time?

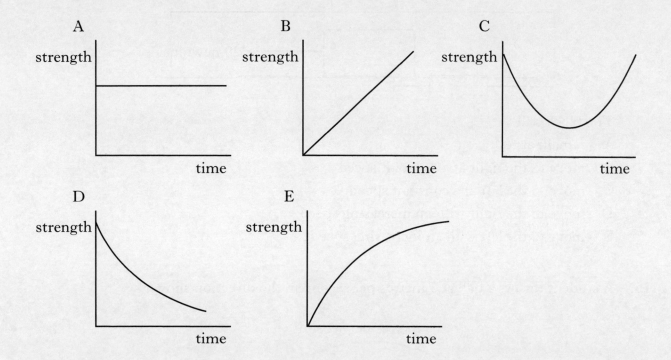

12. During a thunderstorm, the lightning is seen before the thunder is heard.
This is because

A   the eye is more sensitive than the ear

B   the thunder is produced before the lightning

C   the lightning is produced before the thunder

D   light travels faster than sound

E   sound travels faster than light.

13. Which one of the following frequencies of sound can be heard by the average human?

A   4 hertz

B   12 hertz

C   600 hertz

D   25 000 hertz

E   40 000 hertz

14. Sound level is measured in

A   amperes

B   decibels

C   hertz

D   ohms

E   volts.

15. A block is at rest on a smooth surface.

    A force of 20 newtons is now applied to the block as shown.

    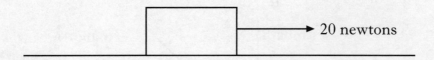

    The block will

    A   remain at rest

    B   move to the right at a constant speed

    C   move to the left at a constant speed

    D   move to the right with an increasing speed

    E   move to the left with an increasing speed.

16. A student throws a ball at 3 metres per second in the direction shown.

    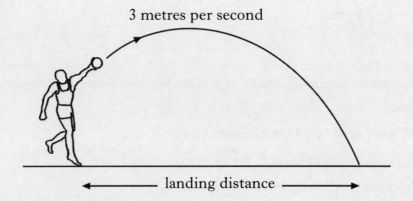

    The landing distance would be increased by

    A   increasing the mass of the ball

    B   increasing the speed of the throw of the ball

    C   increasing the weight of the ball

    D   decreasing the force of the throw

    E   decreasing the speed of the throw of the ball.

17. A car accelerates to overtake a lorry.

Acceleration is the

A   change in speed each second

B   change in time each second

C   change in distance each second

D   change in force each second

E   change in mass each second.

18. A trolley is released from position P on a slope as shown.

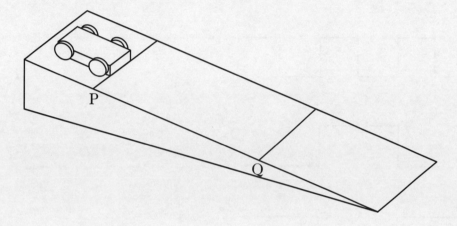

The length of the trolley is 0·2 metres.

To calculate the instantaneous speed of the trolley at Q, we also need to know

A   the distance from position P to position Q

B   the time taken by the trolley to move from position P to position Q

C   the speed of the trolley at position P

D   the time taken for the trolley to pass position P

E   the time taken for the trolley to pass position Q.

19. A student designs an electronic system to sound an alarm when an engine gets too hot.

Which row in the table shows suitable input and output devices for this system?

	Input device	Output device
A	LDR	thermistor
B	buzzer	thermistor
C	thermistor	LDR
D	thermistor	buzzer
E	LDR	buzzer

20. The digital signal shown below is applied to the input of a NOT gate.

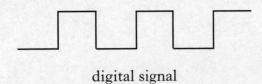

digital signal

The output signal from the NOT gate is

A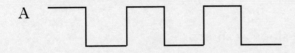

B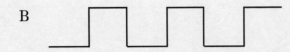

C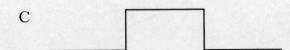

D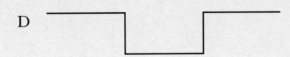

E

Candidates are reminded that the answer sheet for Section A MUST be placed INSIDE the front cover of this answer book.

Marks

SECTION B

Answer questions 21–30 in the spaces provided.

21. Some mobile phones have GPS (Global Positioning System).

This means that you can find out where you are if you are lost.

(*a*) Complete the sentences below using some of these words.

light	**200**	**radio**	**electrical**
sound	**300**	**energy**	**geostationary**

The GPS phone uses _____ waves to receive signals

from satellites in space. These waves transfer _____ .

The waves travel at a speed of _____ million metres per

second. A satellite that stays above the same point on the Earth's surface

is called a _____ satellite.　　　2

(*b*) It takes 0·12 seconds for a signal to travel from a satellite 36 million metres above the Earth to the GPS phone.

   (i) What time does a signal take to travel from a different satellite 18 million metres above the Earth?

                                               1

   (ii) Explain your answer.

                                               1

22. (*a*) A block diagram of a radio receiver is shown.

Marks

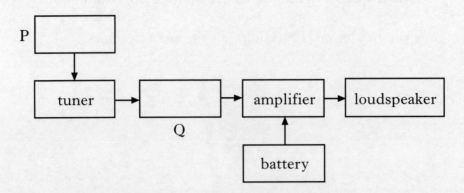

(i) Complete the diagram by labelling blocks P and Q.

2

(ii) What is the function of the tuner?

1

(*b*) Oscilloscopes are connected across the input and output of the amplifier. The settings on each oscilloscope are the same.

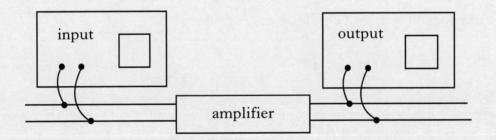

(i) The diagram below shows the input signal.

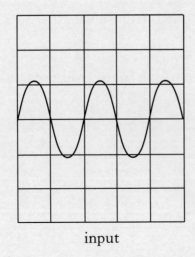

input

output

On the output diagram, draw the output signal from the amplifier.

2

Marks

22. **(b) (continued)**

(ii) The frequency of the input signal is now increased.

What change will there be in the output signal shown on the oscilloscope?

1

(c) The diagram below shows an amplifier connected to a signal generator and a loudspeaker. Voltmeters measure the input and output voltages of the amplifier.

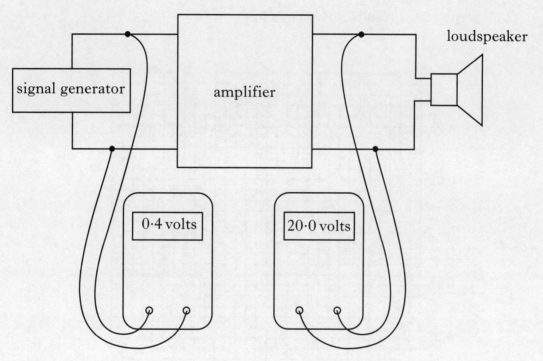

Using information from the diagram, calculate the voltage gain of the amplifier.

2

[Turn over

Marks

23. (a) A technician for a lamp company is investigating the properties of a lamp at different voltages.

For different voltages the technician measures the current in the lamp.

The graph of her results is shown below.

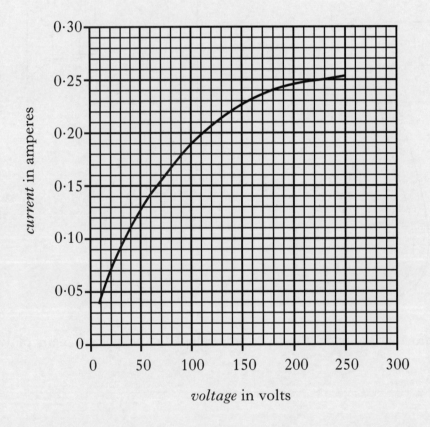

voltage in volts

(i) What is the value of mains voltage?

1

Marks

23. **(*a*)** **(continued)**

(ii) What is the current in the lamp when it is being operated at mains voltage?

1

(iii) Calculate the resistance of the bulb at mains voltage.

2

(*b*) The bulb is now operated at a lower voltage than mains voltage.

(i) Will the resistance of the bulb be **bigger**, **smaller** or **the same as** your answer in (*a*)(iii)?

1

(ii) Explain your answer.

1

[Turn over

Marks

24. The following is part of a crossword.

```
                                    I
                                    N
                                    F
                                    R
                    G  A  M  M  A
                                    R
              X                     E
     U  L  T  R  A  S  O  U  N  D
        A     A
        S     Y
        E
  U  L  T  R  A  V  I  O  L  E  T
```

(a) Select **three** answers from the crossword to complete the table below.

Medical use	Answer from crossword
Scanning an unborn baby	
Detecting broken bones	
Treating vitamin deficiency	

3

(b) What is another name for infrared?

1

Marks

25. (*a*) Laser light reflects in the same way as light from a filament lamp.

Complete the diagram to show the normal and the reflected beam.

2

(*b*) Lasers can be used to cut sheets of metal.

A lens is used to focus the laser light onto the metal.

Complete the following diagram showing the correct shape of lens and the effect it has on the beam of light.

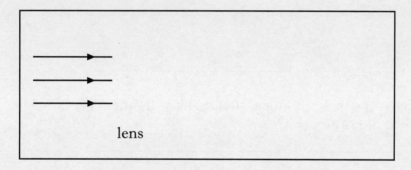

2

(*c*) Some students use a low power laser to try to show the security markings on bank notes. They are not successful.

What type of radiation should they use?

1

[Turn over

Marks

26. (a) A student is learning to play the panpipes. He blows across the pipes and each one produces a different note.

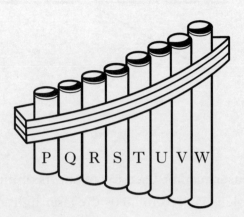

(i) What must the air in the pipes be doing to produce the notes?

1

(ii) Which pipe will produce the highest frequency note when the student blows across it?

1

(iii) The notes from pipes Q and W are an octave apart.

Pipe W produces sound of frequency 256 hertz.

What is the frequency of the sound from pipe Q?

1

Page eighteen

Marks

26. **(continued)**

(*b*) The student investigates the speed of sound in different gases. He designs an experiment as shown below. A short note is produced by the loudspeaker and the sound travels through the tank from one microphone to the other.

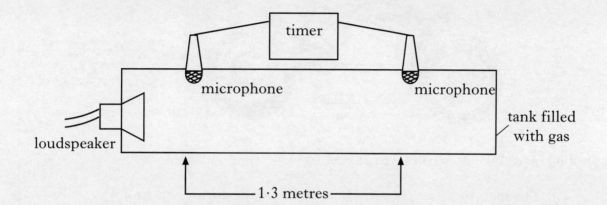

(i) The tank is filled with carbon dioxide. The sound travels a distance of 1·3 metres from the first microphone to the second microphone. The timer records a time of 0·005 seconds.

Calculate the speed of sound in carbon dioxide.

2

(ii) The tank is now filled with a different gas. The time recorded on the timer is less than 0·005 seconds.

(A) Is the speed of sound in this gas **less than**, **equal to** or **more than** the speed of sound in carbon dioxide?

1

(B) Explain your answer.

1

Page nineteen **[Turn over**

Marks

27. The picture shows a motorcyclist on a motorbike.

(*a*) The motorbike goes from 0 to 28 metres per second in 3 seconds.

(i) The motorbike covers a distance of 42 metres during this time. Calculate the average speed.

2

(ii) The motorcyclist is comparing the performance of his motorbike with another bike. The performances of the two bikes are shown in the table.

	Shortest time for 0–28 metres per second in seconds	*Engine power* in brake horse power	*Mass* in kilograms
Bike 1	3	150	181
Bike 2		121	181

Complete the table to show a possible time for Bike 2.

1

Marks

27. (continued)

(*b*) The headlamp of one of the motorbikes uses a 12 volt, 60 watt bulb.

   (i) Calculate the current drawn from the battery when the headlamp is operating.

2

   (ii) Circle the most suitable size of fuse required to protect the bulb circuit.

     3 amperes   10 amperes   13 amperes

1

   (iii) Draw a circuit diagram showing how the bulb, battery, ignition switch and fuse are connected.

2

[Turn over

Marks

28. At the end of a week of skiing lessons, students are given a chance to try a short downhill course.

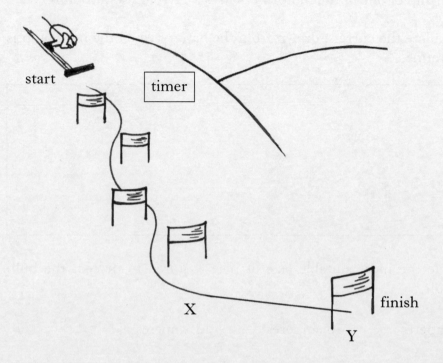

(a) The students want to calculate the average speed of a skier down the course. An electronic timer is used to measure the time between the start and finish lines.

   (i) What electronic device could be used to stop the timer at the finish line?

   1

   (ii) What other measurement is needed to calculate the average speed?

   1

   (iii) How would these measurements be used to calculate the average speed?

   1

DO NOT WRITE IN THIS MARGIN

Marks

28. (continued)

(*b*) (i) The mass of the skier is 60 kilograms.

Calculate the weight of the skier.

2

(ii) Between points X and Y the forces on the skier are balanced.

What happens to the speed of the skier between points X and Y?

1

[Turn over

Marks

29. (*a*) An electronic system consists of three parts.

Complete the block diagram below.

1

(*b*) Some electronic devices are listed below.

microphone   **motor**   **switch**

loudspeaker   **lamp**   **LED**

(i) Complete the table below by putting each device in the correct column.

Input device	*Output device*

3

(ii) What is the energy change in an LED?

1

Marks

29. **(continued)**

(*c*) The resistance of an LDR is measured as the light level increases.
The results are shown on a graph.

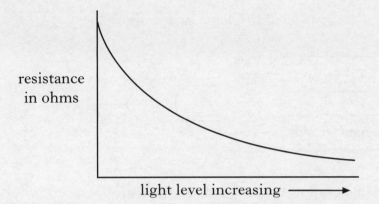

resistance
in ohms

light level increasing ⟶

(i) Looking at the graph, state what happens to the resistance of the LDR as the light level increases.

1

(ii) Name the meter used to measure resistance.

1

[Turn over

Marks

30. A car alarm system has two sensors.

One sensor activates if someone smashes a window.

The second sensor activates if someone tries to move the car.

Each sensor gives a high output when activated.

(*a*) The two sensors are connected to a logic gate as shown.

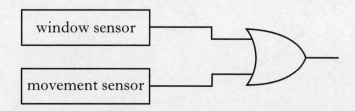

Complete the table to show the output from the logic gate.

Logic level of window sensor	Logic level of movement sensor	Output logic level of gate
0	0	
0	1	
1	0	
1	1	

1

Marks

30. **(continued)**

(b) If the owner wants to drive the car, the alarm must be switched off.

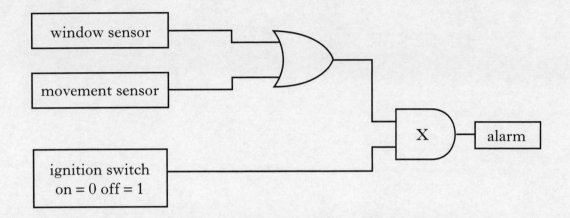

(i) Name logic gate X.

1

(ii) Is the output from the ignition switch **high** or **low** when the switch is on?

1

(iii) Explain why the alarm will not sound if the car is moved when the ignition switch is on.

2

[END OF QUESTION PAPER]

Marks

YOU MAY USE THE SPACE ON THIS PAGE TO REWRITE ANY ANSWER
YOU HAVE DECIDED TO CHANGE IN THE MAIN PART OF THE ANSWER
BOOKLET. TAKE CARE TO WRITE IN CAREFULLY THE APPROPRIATE
QUESTION NUMBER.

[BLANK PAGE]

[BLANK PAGE]

Published by Leckie & Leckie Ltd, 3rd Floor, 4 Queen Street, Edinburgh EH2 1JE
tel: 0131 220 6831, fax: 0131 225 9987, enquiries@leckieandleckie.co.uk, www.leckieandleckie.co.uk

Physics
Intermediate 1 2005

1. E

2. E

3. C

4. B

5. D

6. (a) geostationary

    (b) 24 hours **or** 1 day

    (c) curved reflector (aerial) **or** dish (aerial) **or** sketch

7. (a) (i) radio (signal) **or** microwaves
         (ii) 300 000 000 metres per second

    (b) (i) electrical (signal)
         (ii) (nearly) 300 000 000 metres per second

    (c) speed = $\dfrac{\text{distance}}{\text{time}} = \dfrac{830}{2\cdot5} = 332$ m/s

    (d) (i) light (signal)
         (ii) total internal reflection
         (iii) (nearly) 200 000 000 metres per second

8. (a) (i)

    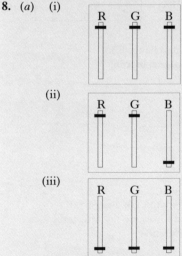

         (ii)

         (iii)

    (b) (i) amplifier
         (ii) to increase strength **or** energy **or** amplitude of signal

9. (a) to cut out **or** trip **or** disconnect circuit if current is too high **or** if there is an overload

    (b) (i) • parallel
            • all have 230 V or independent control or independent failure

9. (b) continued

         (ii) current = $\dfrac{\text{power}}{\text{voltage}} = \dfrac{920}{230} = 4$ (amperes)

    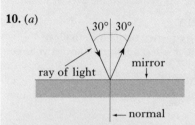

         (iii) resistance = $\dfrac{\text{voltage}}{\text{current}} = \dfrac{230}{4} = 57\cdot5$ ohms

    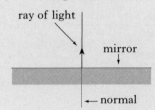

         (iv) • total current = 12 (amperes)
              • the circuit breaker will trip

10. (a)

    (b) back along the normal

    (c) (i) laser parallel or very concentrated/focused or single colour
         (ii) any medical or non-medical use

11. (a)

    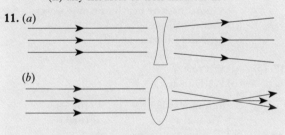

    (b)

    (c) thinner/narrower/less curved

    (d) convex **or** converging **or** correct drawing

12. (a) • Infrared   C **or** G
         • Ultrasound   B **or** C
         • X–rays   D **or** F
         • Ultraviolet   E

    (b) Ultrasound

13. (a) Any one of:
         • distance
         • shielding
         • time
         • other correct answer

Physics
Intermediate 1 2005 (cont.)

13. (b)   (i) 3·5 metres
        (ii) background count **or** there is gamma
           radiation in our surroundings **or** there is
           gamma radiation always around us

14. (a)   (i) 60   office
            105   drilling machine
        (ii) decibel or dB
        (iii) • sound level is very high **or** over 80 dB **or**
            loud
           • this causes damage to the ears or hearing

    (b)   (i) • model Y
           • high frequency signal has smallest
            height/smallest amplitude/lowest volume
        (ii) • model X
           • low frequency signal has smallest
            height/smallest amplitude/lowest volume

15. (a)   (i) 3
        (ii) $\text{average speed} = \dfrac{\text{distance}}{\text{time}} = \dfrac{96}{1} = 96 \text{ m/s}$

    (b) $\text{current} = \dfrac{\text{power}}{\text{voltage}} = \dfrac{12\,000\,000}{25\,000} = 480 \text{ amperes}$

16. (a)   (i) 5·0 kilograms
        (ii) weight = 10 × mass = 10 × 5 = 50 newtons
        (iii) the force **or** pull of the Earth on the object
           or force of gravity
        (iv) newton balance **or** spring balance **or**
           newton meter

    (b)   (i) • more damage
           • more mass **or** heavier **or** more force
        (ii) • more damage
           • greater (impact) speed

17. (a)   (i) OR (gate)
        (ii) AND (gate)

    (b)   • Ignition switch   Q (**or** R)
          • Manual switch   R (**or** Q)
          • Temperature sensor P

    (c)   (i) thermistor
        (ii) LED **or** lamp
        (iii) electrical to kinetic

18. (a)   (i) tape deck **or** CD player
        (ii) loudspeaker (unit)

    (b)   (i) 18 volts
        (ii)
           $\text{voltage gain} = \dfrac{\text{output voltage}}{\text{input voltage}} = \dfrac{18}{0·002} = 9000$

Physics
Intermediate 1 2006

1. E

2. D

3. A

4. C

5. C

6. D

7. E

8. (a) There are no wires between transmitter and
       receiver
       OR
       signals received by radio waves to the aerial.

    (b) Lower

    (c) 300 million metres per second.

    (d)

Transmitter	*Frequency* in millions of hertz
X	89·4
W	94·3
Z	94·3
Y	98·2

    X and Y are interchangeable
    W and Z are interchangeable

9. (a)

    (b) Copper wire network OR network B
       because optical fibres are not affected by
       electrical interference
       OR
       copper conducts electric current.

    (c) optical fibres are difficult to join together.

10. (a)   (i) $\text{current} = \dfrac{\text{power}}{\text{voltage}} = \dfrac{805}{230} = 3·5 \text{ amperes}$

        (ii) 5 amperes

10. (b) (i)

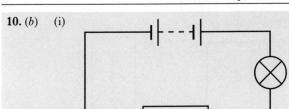

     (ii) open circuit

  (c) infrared (radiation)

11. (a) (i) switches S_1 and S_2
     (ii) electrical (energy) to kinetic (energy)

  (b) (i) 230 volts
     (ii) 230 volts

12. (a) variable resistor

  (b) resistance $= \dfrac{\text{voltage}}{\text{current}} = \dfrac{4}{0 \cdot 5} = 8$ ohms

  (c) Add more batteries OR increase the voltage. Decrease resistance of variable resistor OR remove variable resistor.

13. (a) reflected

  (b)

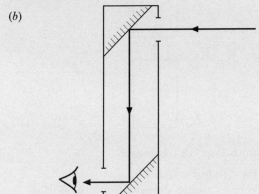

  (c) The mouthpiece of the mobile phone contains a **microphone**. This device changes **sound** energy to **electrical** energy. An **LED** is used to light up the keypad. This device changes **electrical** energy to **light** energy.

14.

Radiation	Medical Uses	Non-medical uses
X-Rays	detecting broken bones	scanning baggage at airports
Gamma	sterilising surgical instruments	detecting leaks from cracked pipes
Infrared	treating muscle strain	night vision binoculars
Ultraviolet	treating acne	checking security markings on bank notes

15. (a) (i) carries light into the patient OR into the windpipe
     (ii) carries light out of the patient OR allows doctor to see the windpipe

15. (b)

  (c) laser light has one colour OR laser light is concentrated

16. (a) (i) R
     (ii) Q

  (b) 20 hertz to 20 000 hertz

  (c) speed $= \dfrac{\text{distance}}{\text{time}} = \dfrac{1 \cdot 3}{0 \cdot 004} = 325$ metres per second

  (d) we hear our own voice (by vibrations) in bones and air
    we hear a recording (by vibrations) in air only

17. (a) weight $= 10 \times$ mass $= 10 \times 65 = 650$ newtons

  (b) (i) 5 metres per second
     (ii) 10 metres per second
     (iii) acceleration

  (c) she takes less time because she is accelerating

  (d) friction due to the water (is greater than friction due to the air)

18. (a) (i) distance (from cymbals to timer)
        time (between seeing cymbals crash and hearing the sound)
     (ii) measuring tape
        stop watch
     (iii) speed = distance / time

  (b) reaction time OR wind

19. (a) forces are balanced OR the forces are equal and opposite

  (b) average speed $= \dfrac{\text{distance}}{\text{time}} = \dfrac{8100}{15 \times 60}$

    $= 9$ metres per second

  (c) car may have had to stop on the journey OR any other suitable answer involving variable speed on the journey

20. (a) (i) AND gate
     (ii) buzzer OR LED OR motor
     (iii) switch OR LDR
     (iv)

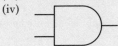

  (b) voltage gain $= \dfrac{\text{output voltage}}{\text{input voltage}} = \dfrac{20}{0 \cdot 05} = 400$

21. (a) (i) OR (gate)
     (ii) NOT (gate)

  (b) (i) (logic) 1
     (ii)

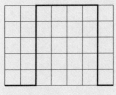

Output C

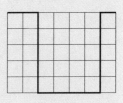

Output D

Physics
Intermediate 1 2007

SECTION A

1. B	**2.** C	**3.** B	**4.** C
5. D	**6.** A	**7.** A	**8.** D
9. C	**10.** B	**11.** B	**12.** B
13. E	**14.** A	**15.** D	**16.** D
17. E	**18.** D	**19.** E	**20.** E

SECTION B

21. (*a*) Radio and TV

    (*b*) microphone

    (*c*)   (i) loudspeaker
          (ii) vibrate OR turn electrical energy into
              sound energy

    (*d*)   (i) to make the signal stronger OR
              to direct the signals to the aerial

          (ii)

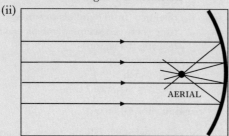

          (iii)

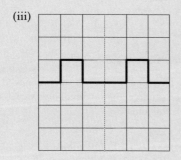

22. (*a*)   (i) series
          (ii) headlights can only come on when the
              ignition is on
          (iii) 9 amperes

    (*b*)   (i) 9·6 volts
          (ii) 4 ohms
          (iii) A (speed will) decrease
              B current is smaller

23. (*a*) speed of sound less than speed of TV signals

    (*b*)

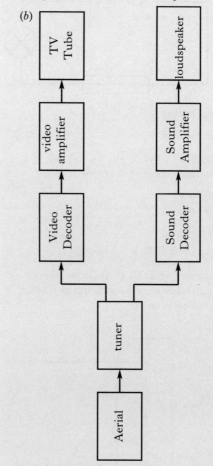

24. (*a*) convex OR converging

    (*b*)   (i) less bright
              smaller diameter (collects less light)
          (ii) smaller
              thinner lens OR lens magnifies less

25.

Radiation	
Gamma	E
X-rays	B
Ultra-violet	F
Visible light	A
Infra red	D
Radio waves	C

26. (*a*)   (i)

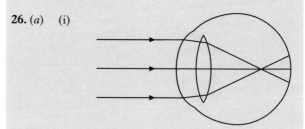

26. (*a*) (ii)

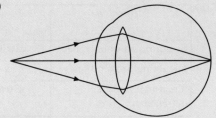

(*b*) concave OR diverging

27. (*a*) transmitter
receiver

(*b*) (i) signal has lost energy
(ii) 0·2 seconds
(iii) 1530 metres per second

28. (*a*) 750 newtons

(*b*) to change or increase speed

(*c*) (i) air resistance OR friction force OR drag

weight OR gravity force

(ii) stays the same

(*d*) (i) (speed) decreases
(ii) (upward force) increases

29. (*a*) ball R

(*b*) 1. Measure the width of the ball.
2. Find the time the ball takes to pass
through the light gate (i.e. time on timer)
3. speed = width of ball ÷ time on timer

30. (*a*) (i) AND gate
(ii)

A	B	C	D
0	0	1	0
0	1	1	1
1	0	0	0
1	1	0	0

(*b*) (i) 175

(ii) 4 amperes

31. (*a*) input device

(*b*) 1. Heat water with bunsen burner.
2. Measure resistance of thermistor with
ohmmeter.
3. Measure temperature of water with
thermometer.
4. Take readings for several different
temperatures.

(*c*) (i) not working correctly
(ii) The resistance did not change as the
temperature increased.

Physics
Intermediate 1 2008

SECTION A

1. B	11. D
2. B	12. D
3. C	13. C
4. B	14. B
5. C	15. D
6. E	16. B
7. E	17. A
8. B	18. E
9. E	19. D
10. C	20. A

SECTION B

21. (a) The GPS phone uses <u>radio</u> waves to receive
signals from satellites in space. These waves
transfer <u>energy</u>
The waves travel at a speed of <u>300</u> million
metres per second. A satellite which stays above
the same point on the Earth's surface is called a
<u>geostationary</u> satellite.

   (b) (i) 0·06 seconds
      (ii) <u>Distance</u> is halved (so time will be halved)

22. (a) (i) P: Aerial
          Q: Decoder
     (ii) To select one signal/station/frequency
         (from many)

   (b) (i)

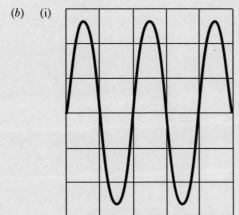

      (ii) There will be more waves (on the screen.)

   (c) 50

23. (a) (i) 230 volts
      (ii) 0·25 amperes
     (iii) 920 ohms

   (b) (i) Smaller
      (ii) By calculation eg $\frac{100}{0·19}$ = 526 ohms

24. (a)

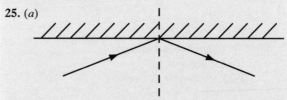

Medical use	Answer from crossword
Scanning an unborn baby	**ultrasound**
Detecting broken bones	**x-ray**
Treating vitamin deficiency	**ultraviolet**

   (b) Heat (radiation)

25. (a)

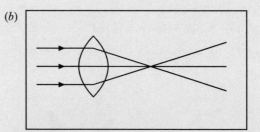

   (b)

   (c) Ultraviolet

26. (a) (i) Vibrate/Vibrating
      (ii) P
     (iii) 512 hertz

   (b) (i) 260 metres per second
      (ii) (A) More than
          (B) Less time for the same <u>distance</u>.

           OR

        e.g. $\frac{1·3}{0·002}$ = 650 metres per second

27. (a) (i) 14 metres per second

      (ii) any time > 3 seconds

   (b) (i) 5 amperes
      (ii) 10 amperes
     (iii)

28. (a) (i) Light gate
      (ii) Distance
     (iii) average speed = $\frac{distance}{time}$

   (b) (i) 600 newtons
      (ii) Stays the same/steady/constant

29. (*a*)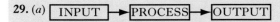

| | INPUT | → | PROCESS | → | OUTPUT |

(*b*) (i)

Input Device	Output Device
microphone	loudspeaker
switch	lamp
	motor
	LED

(ii) Electrical to light

(*c*) (i) (resistance) decreases
(ii) Ohmmeter

30. (*a*)

Logic level of window sensor	Logic level of movement sensor	Output logic level of gate
0	0	**0**
0	1	**1**
1	0	**1**
1	1	**1**

(*b*) (i) AND gate
(ii) low
(iii) the AND gate requires two high inputs (to activate the alarm) if <u>ignition</u> switch is on then one of the inputs is low

[BLANK PAGE]

[BLANK PAGE]